SAAB 29B Tunan

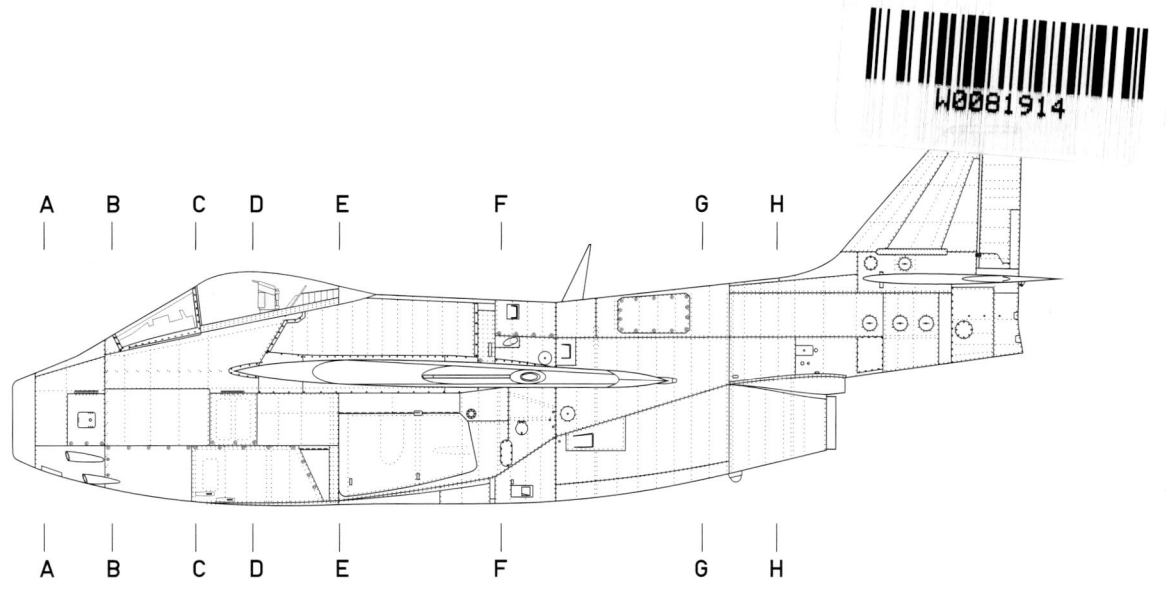

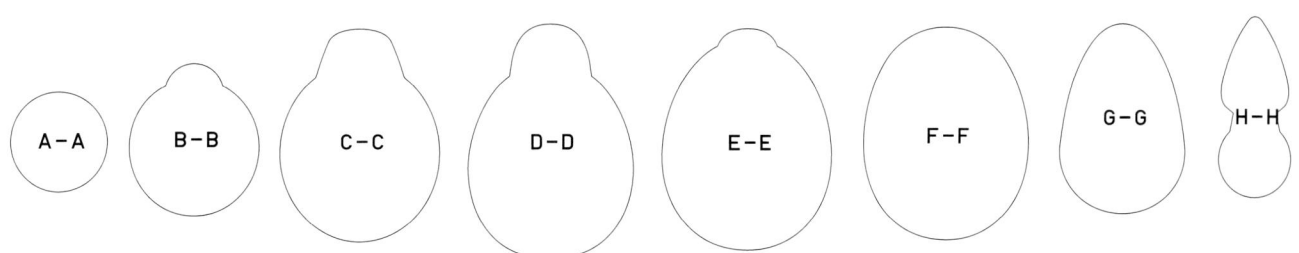

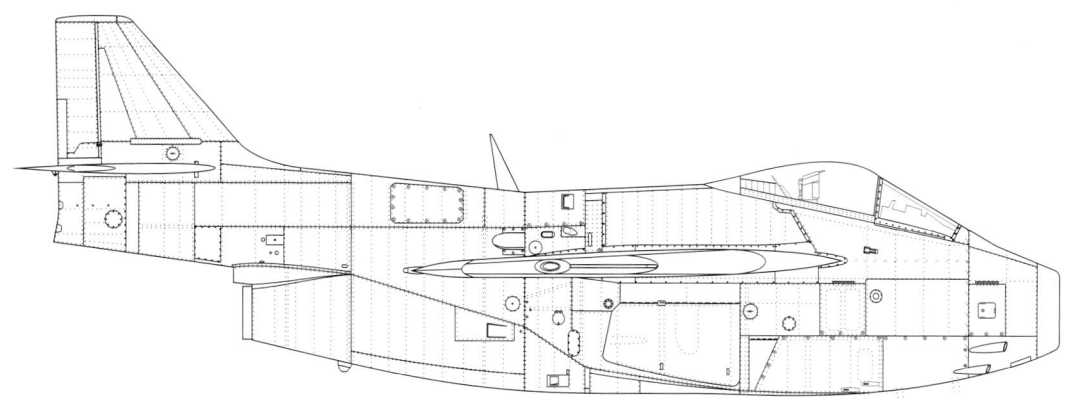

1/72

Drawings: Dariusz Karnas

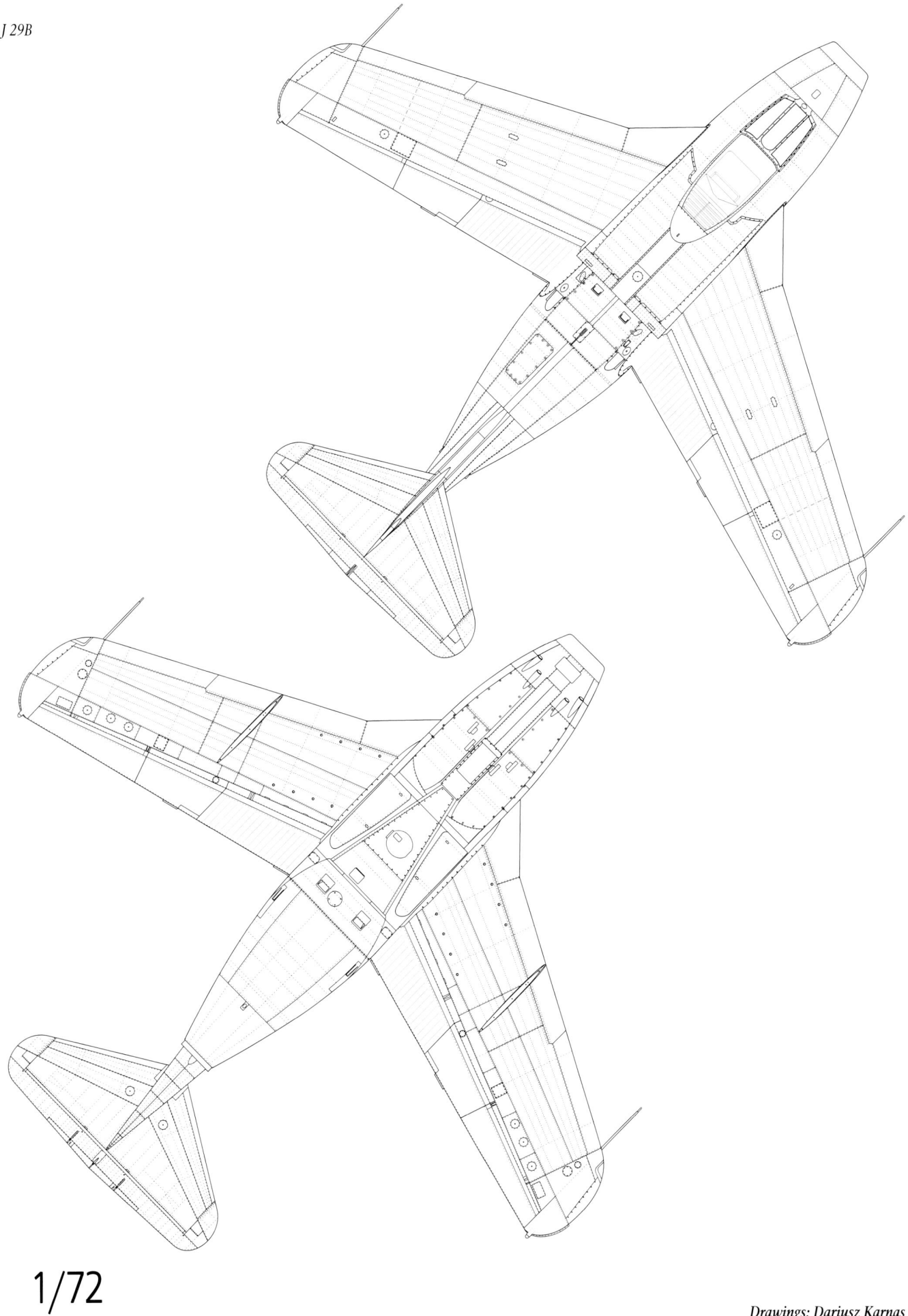

1/72

Drawings: Dariusz Karnas

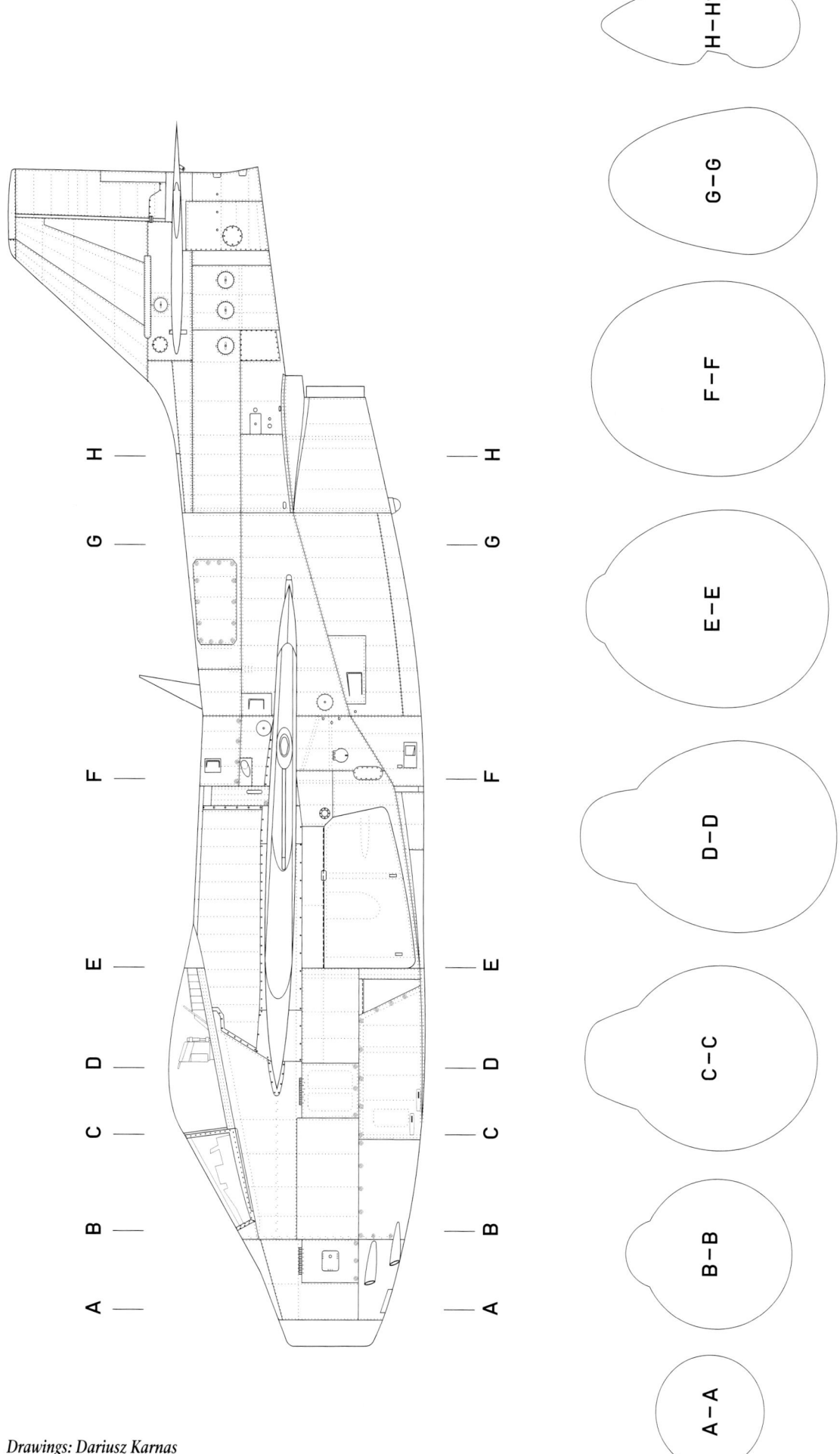

J 29B

H–H

G–G

F–F

E–E

D–D

C–C

B–B

A–A

1/48

Drawings: Dariusz Karnas

3

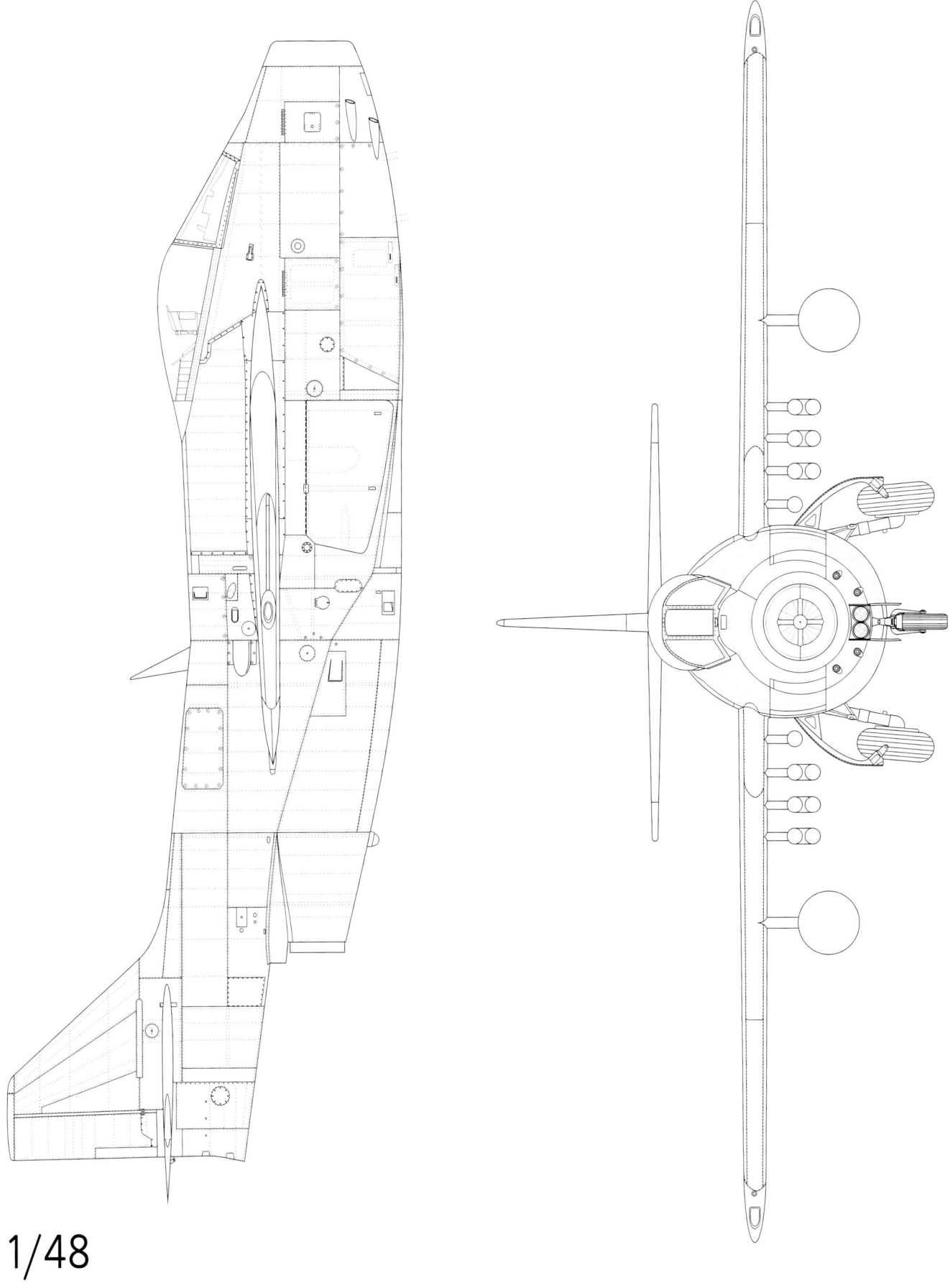

1/48

Drawings: Dariusz Karnas

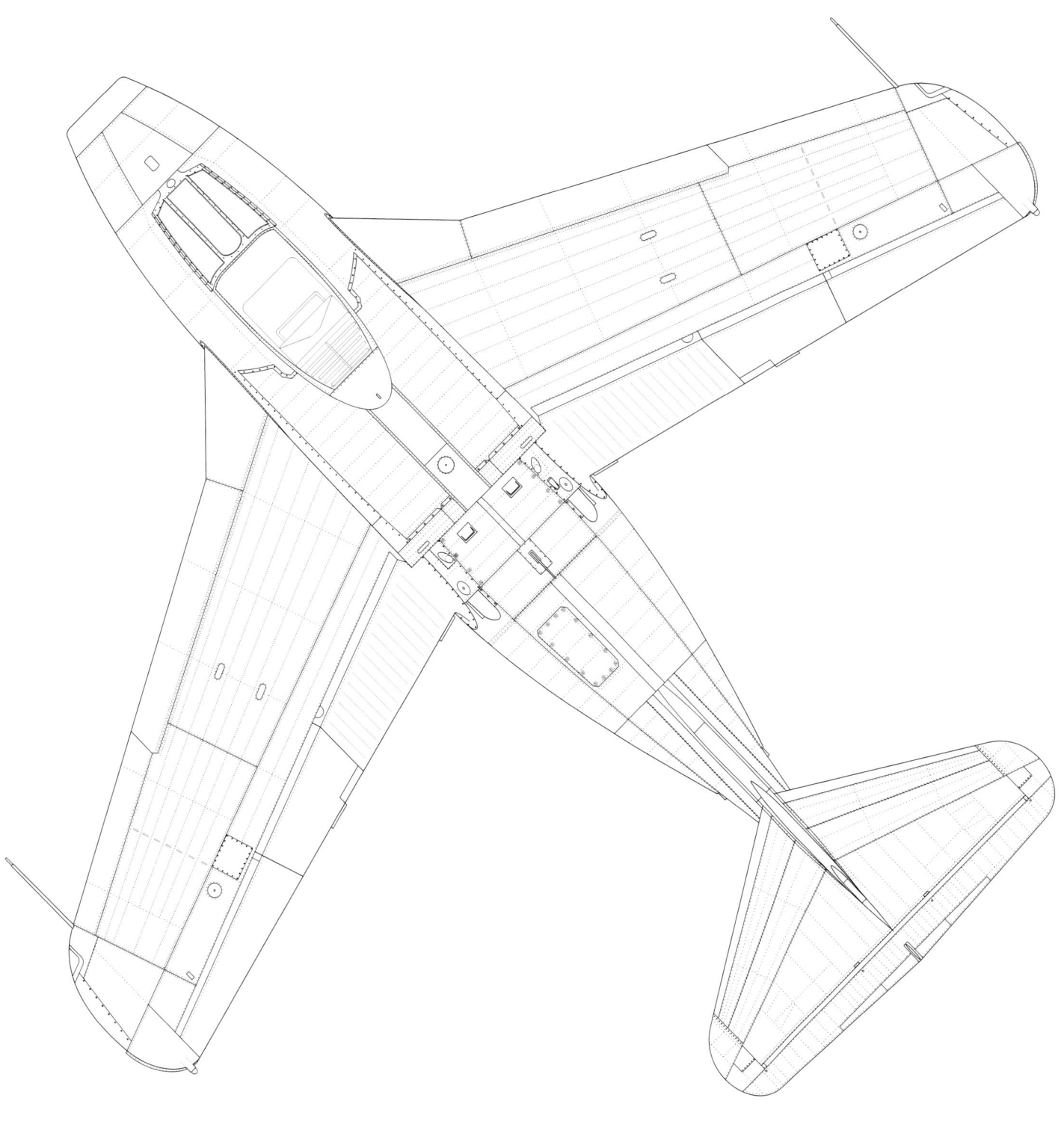

1/48

Drawings: Dariusz Karnas

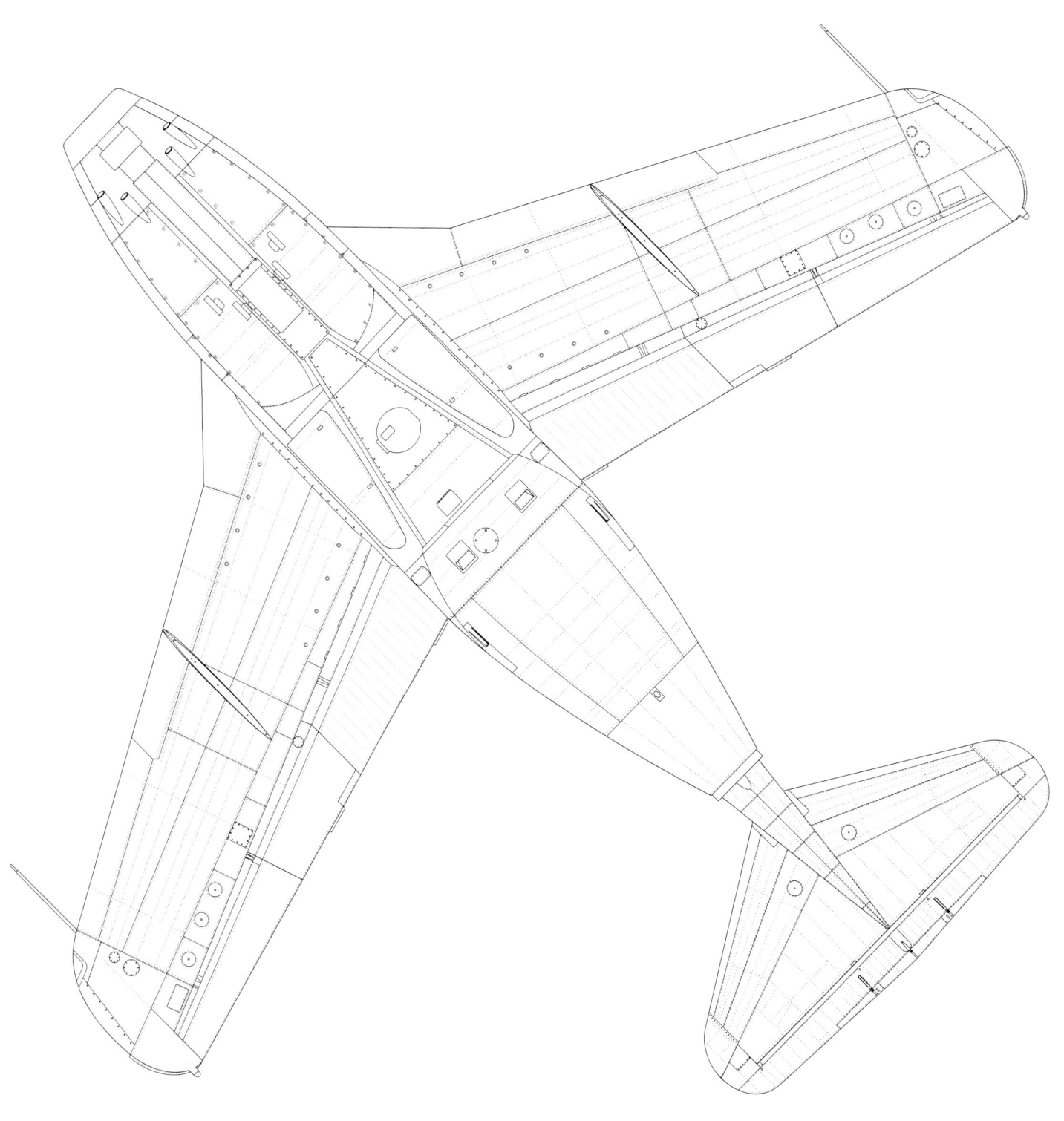

1/48

Drawings: Dariusz Karnas

J 29B, s/n 29484, F 3 blue I, during 1954. (Copyright SAAB AB)

SAAB J 29B, s/n 29386, Yellow E, No. 3 Squadron Wing F 8 participating in air pentathlon, hence the special markings consisting of white numerals on black triangles. The code letters are yellow, outlined in black. (Archives of Svensk Flyghistorisk Förening)

A J 29B of Wing F 16, most likely code Red B outlined in white. Note the red nose ring and that the front end of the drop tank is painted red. (Archives of Svensk Flyghistorisk Förening)

Sergeant Hultqvist about to enter J 29B code Red (?) C of Wing F 8. (via Wing F 8 Association)

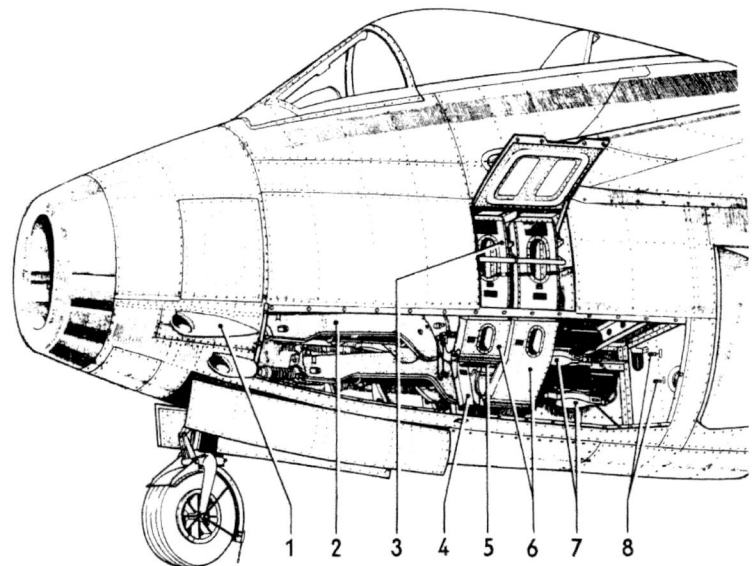

General view of the 20 mm m/47 cannon in the nose of the aircraft. Drawing taken from Technical Manual.

The only SAAB J 29B that has been preserved, s/n 29398, seen here at Flygvapenmuseum in 1989. It is coded White F of Wing F 22. (Mikael Forslund)

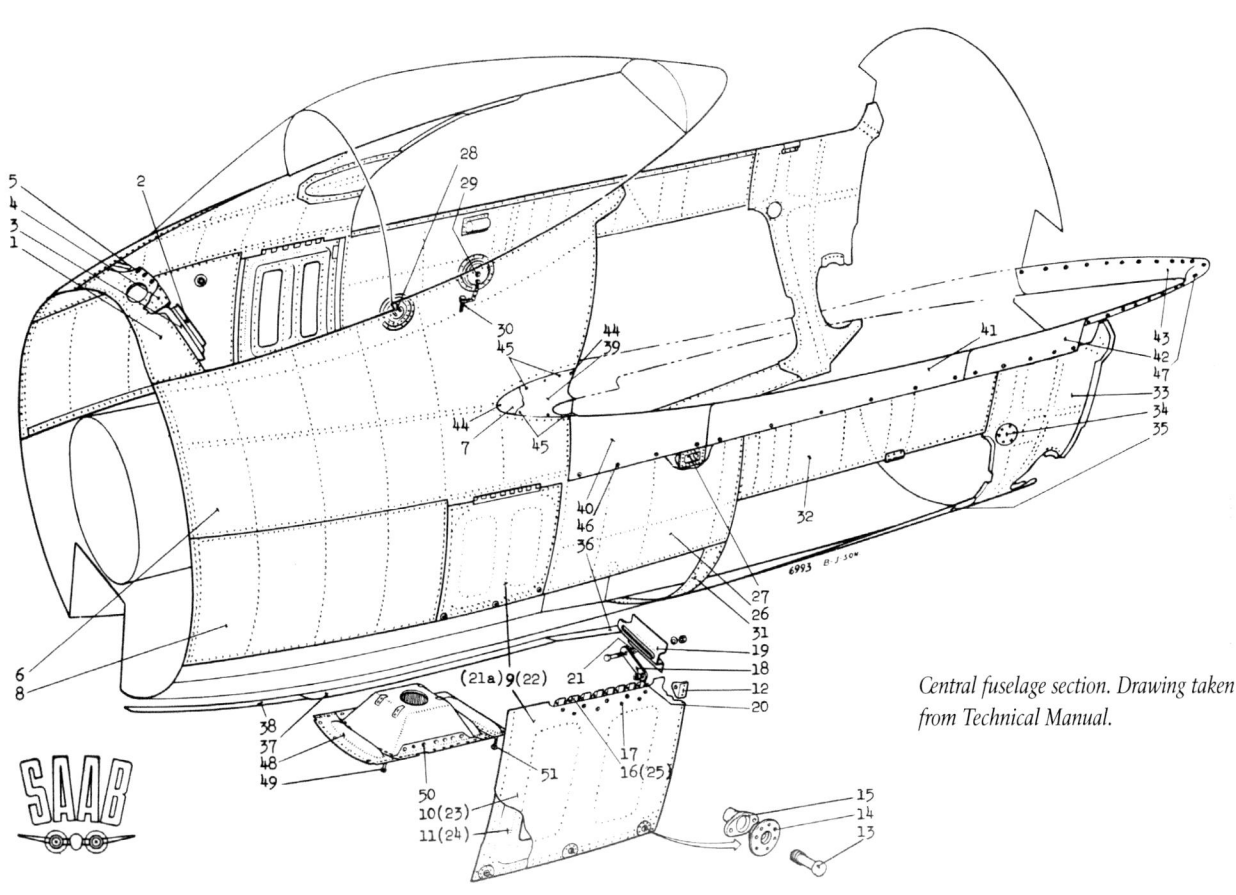

Central fuselage section. Drawing taken from Technical Manual.

Rear fuselage section. Drawing taken from Technical Manual.

Tillv-skylt

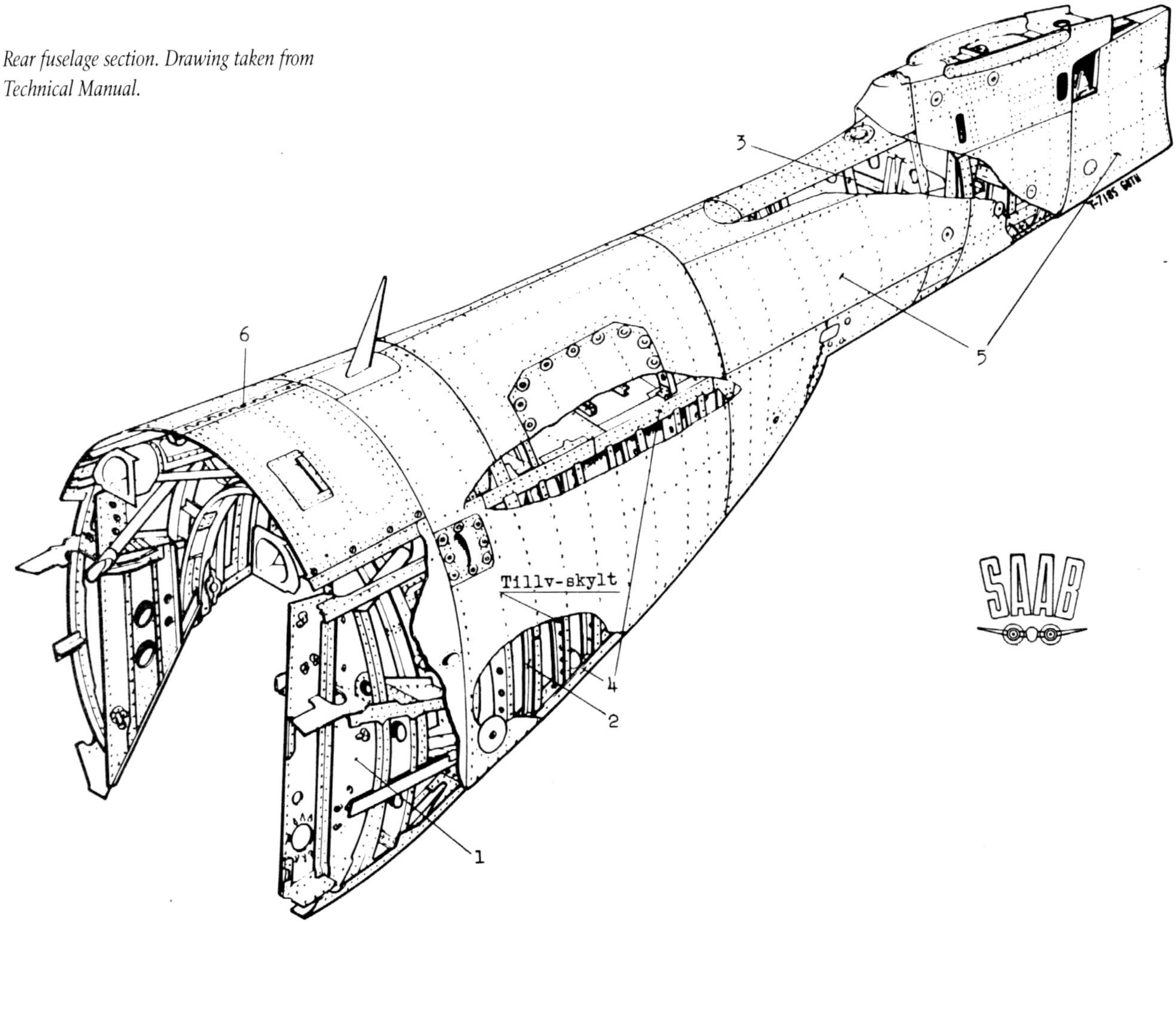

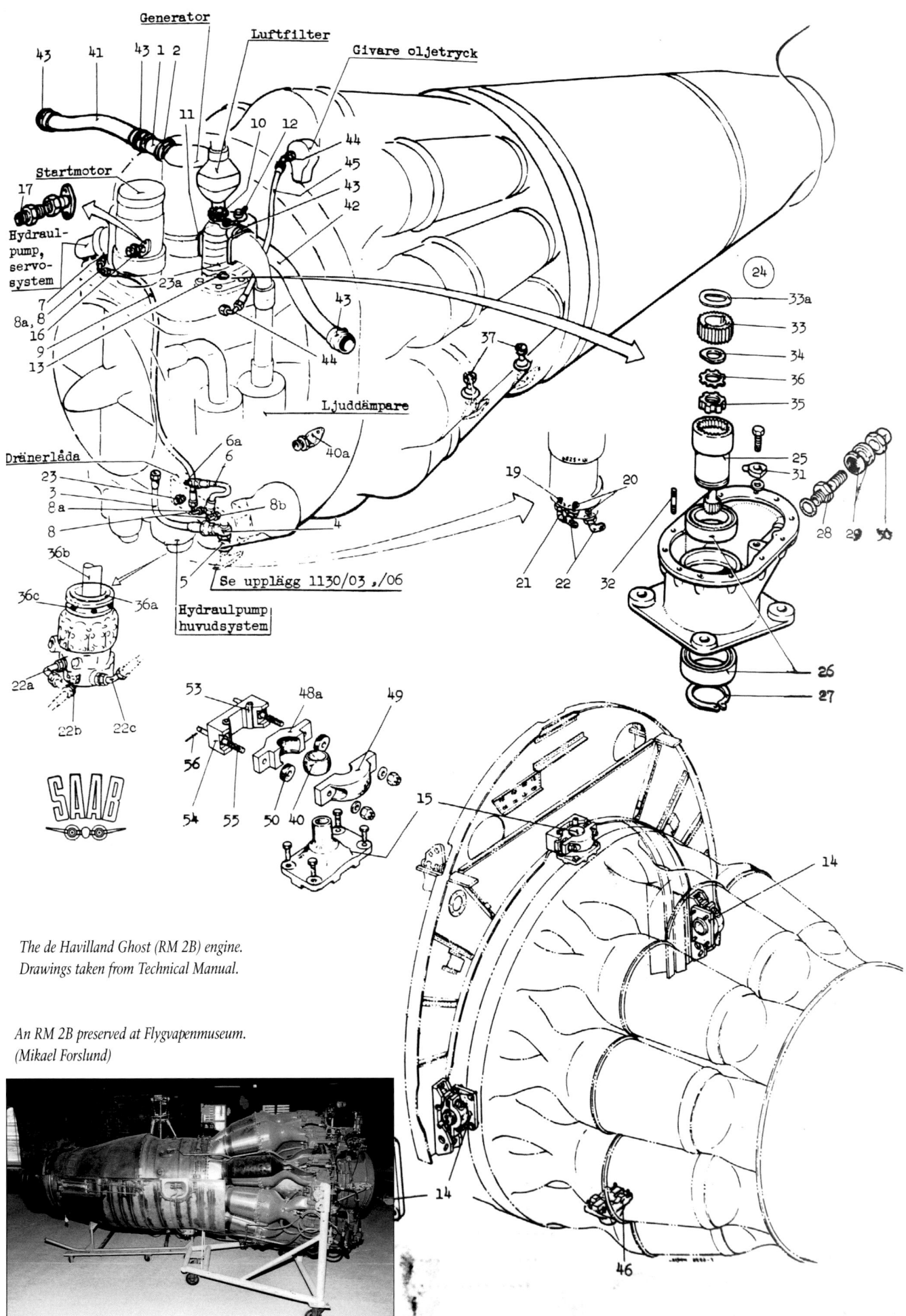

Generator

Luftfilter

Givare oljetryck

Startmotor

Hydraul-
pump,
servo-
system

Ljuddämpare

Dränerlåda

Se upplägg 1130/03 ,/06

Hydraulpump
huvudsystem

SAAB

The de Havilland Ghost (RM 2B) engine.
Drawings taken from Technical Manual.

An RM 2B preserved at Flygvapenmuseum.
(Mikael Forslund)

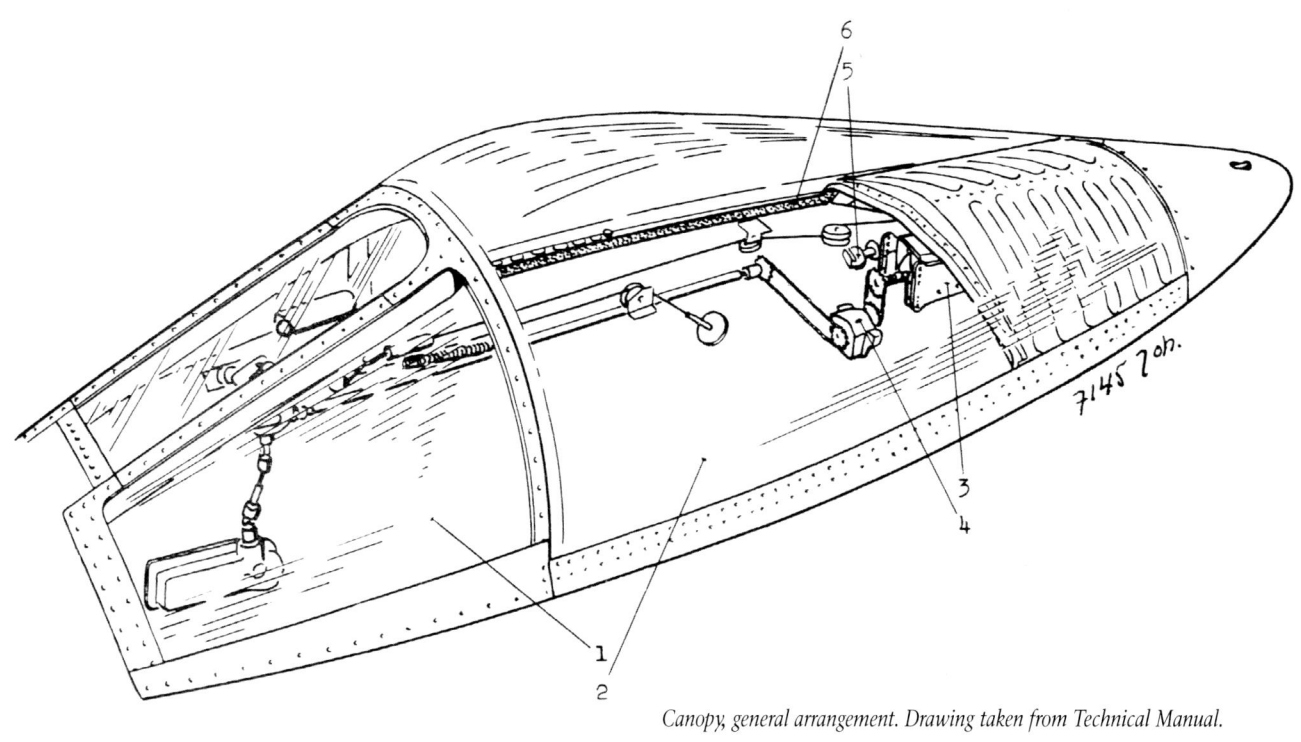

Canopy, general arrangement. Drawing taken from Technical Manual.

A J 29B-pilot of Wing F 22 gets ready for take-off. The photo was taken in 1963. (via Archives of Swedish Aviation Historical Society)

SAAB 29 cockpit, port side. (Technical Manual)

1. Flap lever
2. Circuit closer selector for automatic cannon, 12.7 mm gunnery practice installation
3. Circuit closer for stop valve
4. Map light
5. Tap for securing the cannons before landing
6. HT-fuel lever
7. Red light
8. Throttle with turning handle for distance adjustment of gyro gunsight and button for release of sight and starting the cameras
9. Button for emergency stop of camera
10. Circuit breaker for afterburner-LT-fuel tap
11. Switch release handle for starter rockets (not operational)
12. Circuit breaker for landing light
13. Emergency tap for auxiliary static pressure
14. Release handle for rocket racks
15. Main circuit breaker
16. Manoeuvring box for gyro gun sight
17. Emergency handle, undercarriage down
18. Hydraulic pressure gauge
19. Emergency air pressure gauge
20. Outer air pressure gauge
21. Release button for fire extinguisher

22. LT-fuel handle lock
23. LT-fuel handle
24. Tailplane adjustment indicator
25. Button, "Undercarriage up"
26. Rudder trim indicator
27. Retraction button lock, undercarriage
28. Rudder trim wheel
29. Button, "Undercarriage down"
30. Emergency circuit breaker for tailplane adjustment
31. Throttle friction brake
32. Aileron trim indicator
33. Aileron trim wheel
34. Disconnecting crank for control column centering
35. Mechanical indicator, main undercarriage
36. G-Suit valve
37. Air brakes emergency air valve
38. Flap emergency air valve
39. Flight log compartment
40. Drop tank release lever
41. Air brake lever
42. Volume control lever, missile
43. Emergency firing switch, missile

This and opposite page:
Cockpit – general view. Drawing and photo taken from Technical Manual.

1. Horizontal beam focus wheel
2. Bank index switch gear
3. Artificial horizon indicator
3a. Horizontal beam trim wheel
4. Volume control for horizontal beam
5. Gyro sight
6. Deviation correction card for remote compass
7. Course indicator
8. Release button for artificial horizon
9. "Free gyro" indicator
9a. Course setting device
10. "Locked gyro" indicator
11. Artificial horizon lock
12. Fire warning lights
13. Clock
14. RPM indicator
15. NOZZLE warning indicator
16. Warm air tap for de-icing the windscreen
17. Fuel capacity indicator

18. Nozzle temperature indicator
19. Windscreen electrical heating switch
20. Fuel gauge selector
21. LOW FUEL warning indicator
22. Direction finding indicator
23. Rear bearing thermometer
24. Map compartment
25. Fresh air lever
26. Cabin heating lever
27. Nose wheel mechanical indicator
28. Cabin heating lever
29. Nose wheel steering lever
30. Rudder locking flap
31. Wheel brake parking handle
32. Undercarriage indicator lights
33. Aileron servo release handle
34. R/T button
35. Tailplane incidence adjustment switch
36. Cannon firing button

37. Safety catch
38. Rocket firing button (alt. electrical release of drop tanks)
39. Turn-and-bank indicator
40. Cabin pressure gauge
41. Altimeter
42. Variometer
43. Compass light switch
44. Auxiliary compass
45. Accelerometer
46. Mach-indicator
47. UNDERCARRIAGE NOT LOWERED warning indicator
48. UV-light
49. Auxiliary artificial horizon
50. UHF-dampener

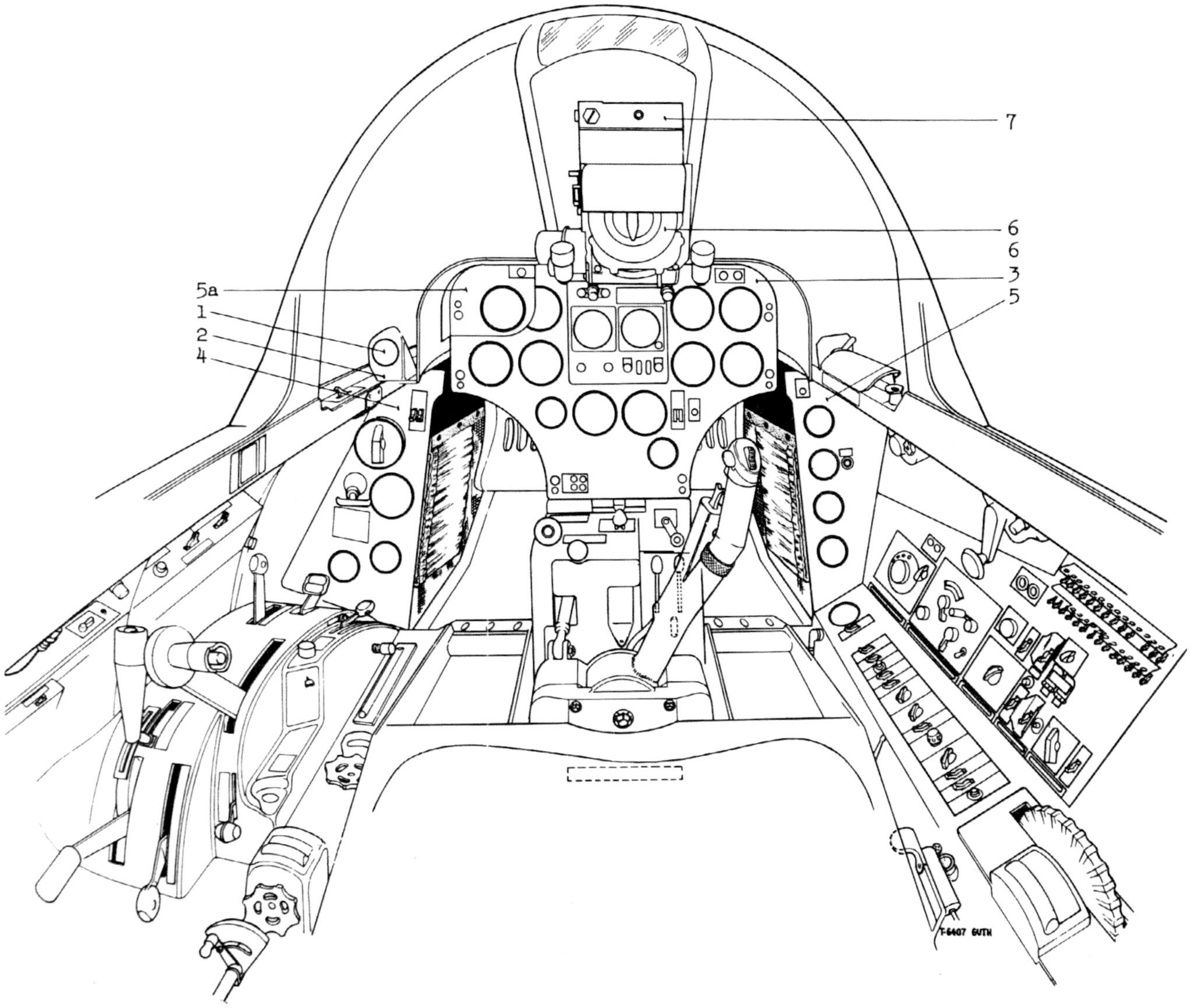

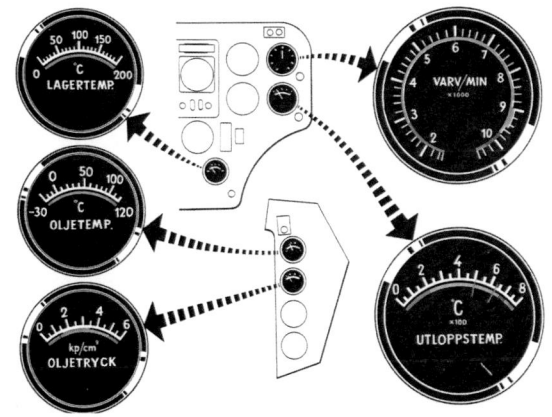

7

6
6
3
5

5a
1
2
4

Hydraulic pressure gauge.

LAGERTEMP. – Bearing temp.

VARV/MIN – Rev/Min.

OLJETEMP. – Oil temp.

UTLOPPSTEMP. – Nozzle temp.

OLJETRYCK – Oil Pressure

Compressed air pressure gauge.
Drawings taken from Technical
Manual.

15

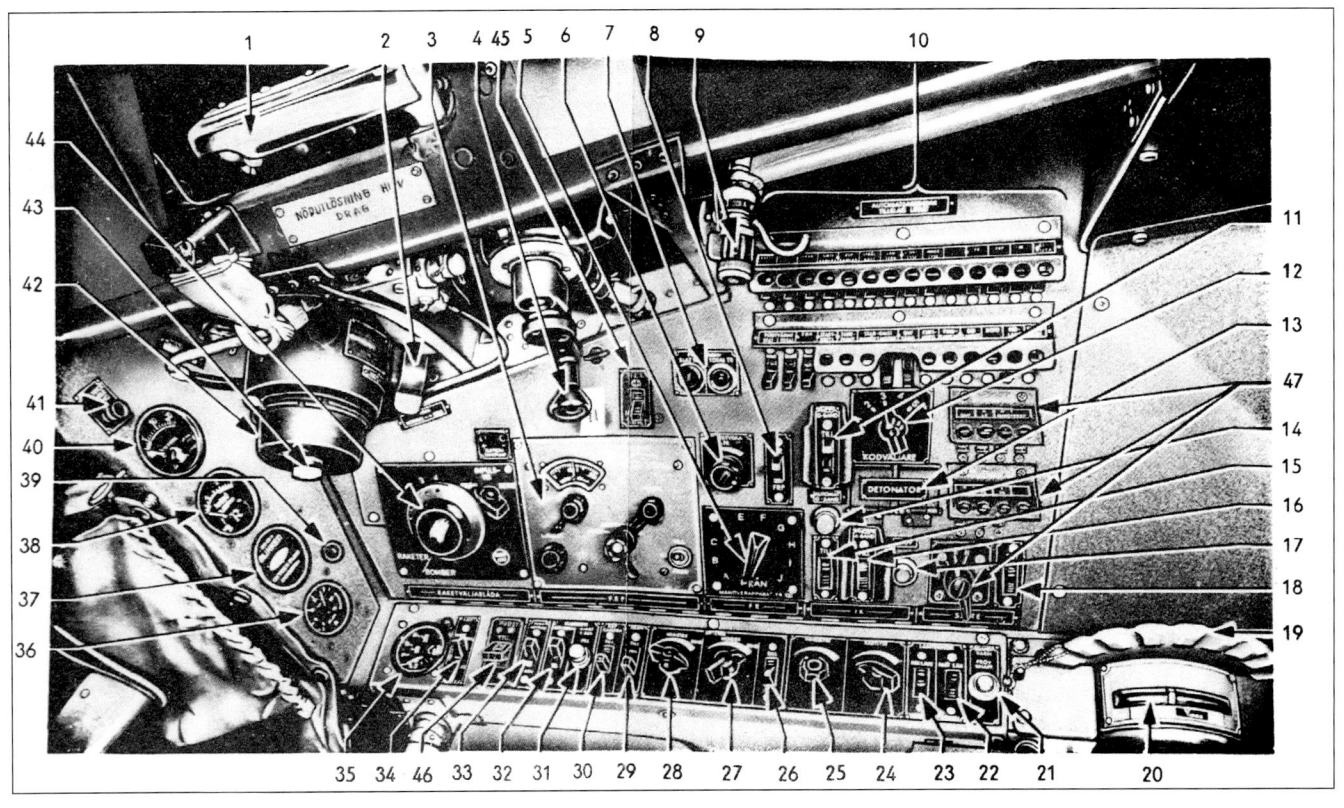

SAAB 29 cockpit, starboard side. (Technical Manual)

1. Canopy emergency release handle
2. Box switch for oxygen regulator
3. Manoeuvring box for radio direction finder
4. Canopy closing and opening handle
5. Radio channel selector
6. Radio volume control
7. "Pitot tube Heating On" indicators
8. (FR-FRP)
9. Red light
10. Automatic fuses
11. Switch for emergency signalling on G-band
12. Code selector
13. Detonator for IK
14. Button for AUTOMATIC G-BAND
15. Switch for turning on and off of IK
16. Switch for emergency signalling on A-band
17. Test button
18. Switch for turning on and off gun sight
19. Emergency elevator trim wheel
20. Emergency elevator trim indicator
21. Fire warning indicator test button
22. Navigational light circuit breakers
23. Navigational light circuit breakers
24. Rheostat for map light

25. Rheostat for red light
26. Switch for UV-light
27. Rheostat for UV-light, right
28. Rheostat for UV-light, left
29. Switch for firing circuit, automatic cannons
30. Switch for gun camera
31. Starter button
32. Manoeuvring switch for starter system
33. Switch, wing fuel tanks
34. Switch for ROCKET FIRING – RELEASE OF DROP TANKS – ALT MANOEUVRING CURRENT – CAMERA POD
35. Volt-ammeter
36. Oxygen pressure gauge
37. Oxygen indicator
38. Oil pressure gauge
39. Generator warning indicator
40. Oil thermometer
41. Low fuel pressure warning indicator
42. Oxygen regulator
43. Oxygen mask sealing, test button
44. Rocket selector box
45. Auxiliary artificial horizon selector switch
46. Missile selector switch

16

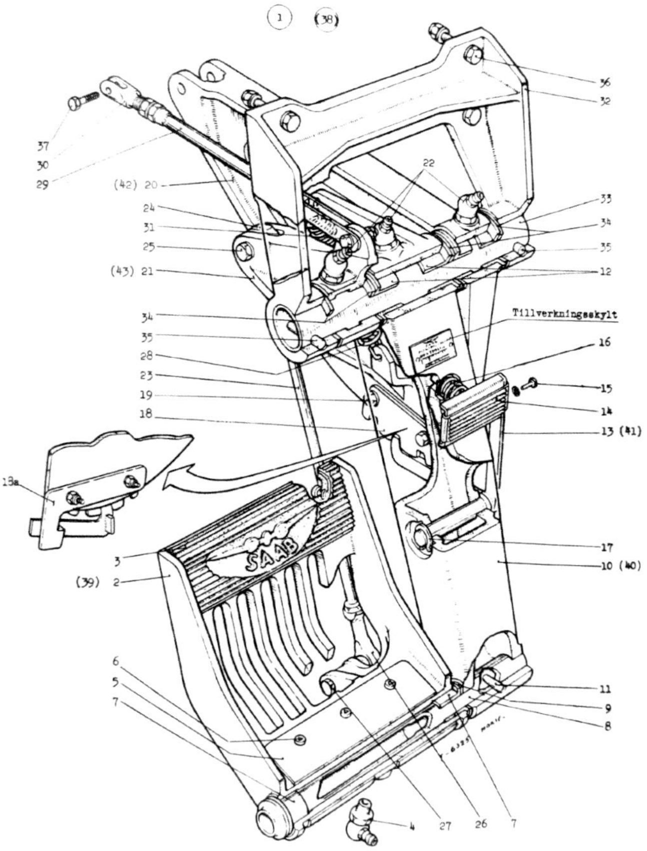

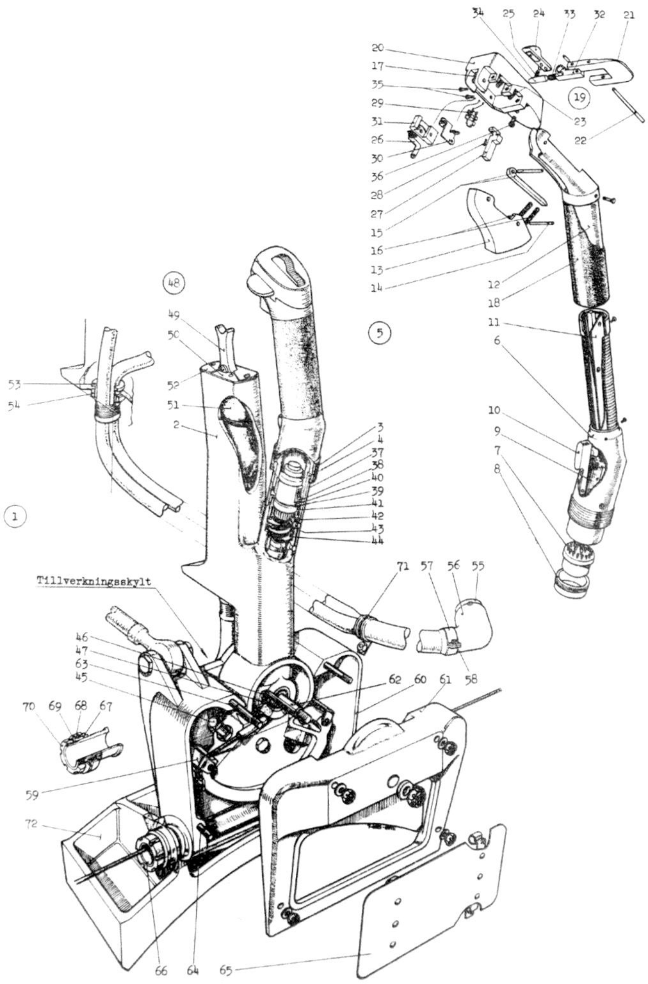

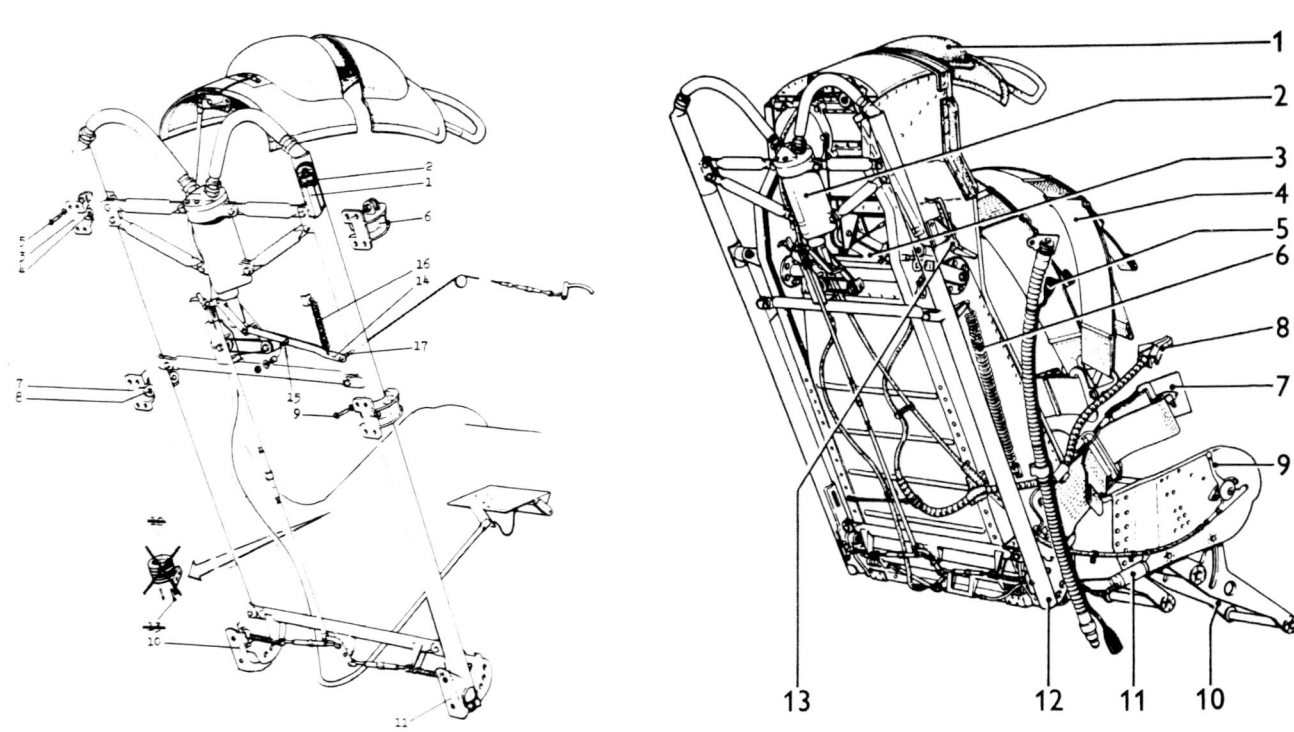

The stick (right) and pedal (left). Drawings taken from Technical Manual.

Pilot seat installation (left) and instalation with the pilot seat (right). Drawings taken from Technical Manual.

17

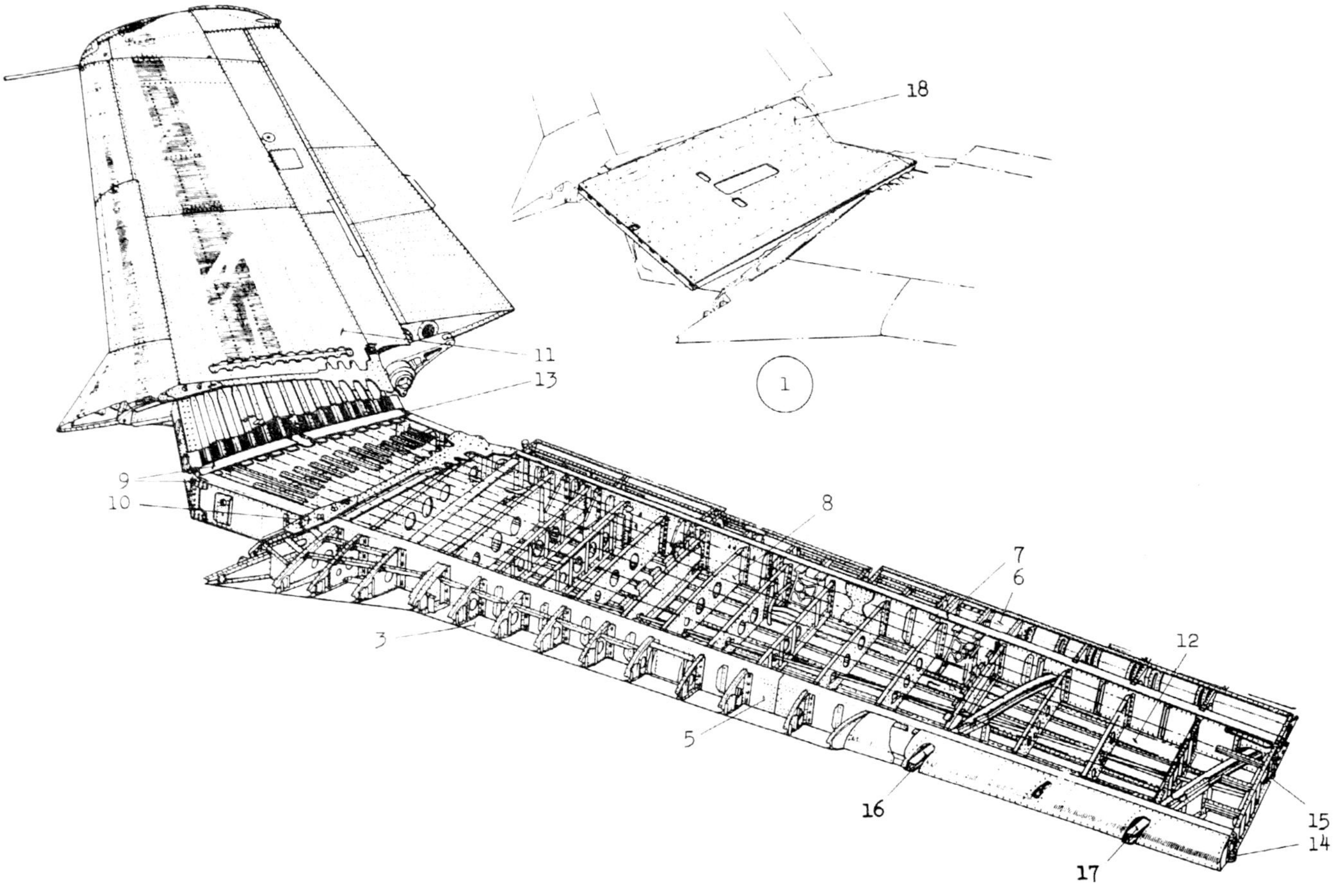

Wings structure. Drawing taken from Technical Manual.

SAAB J 29B code White E (s/n 29393) above Congo. (via Archives of Swedish Aviation Historical Society)

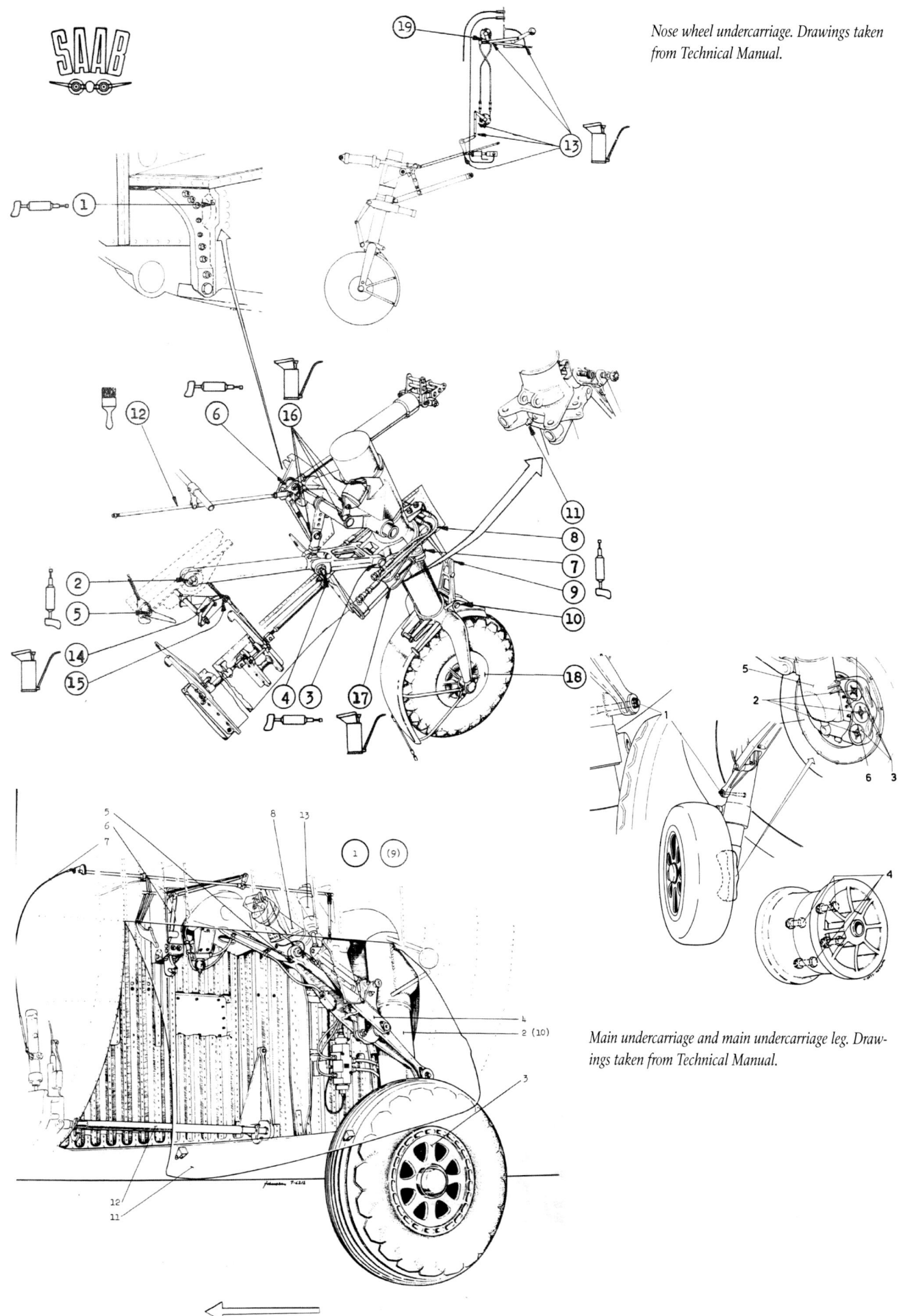

Nose wheel undercarriage. Drawings taken from Technical Manual.

Main undercarriage and main undercarriage leg. Drawings taken from Technical Manual.

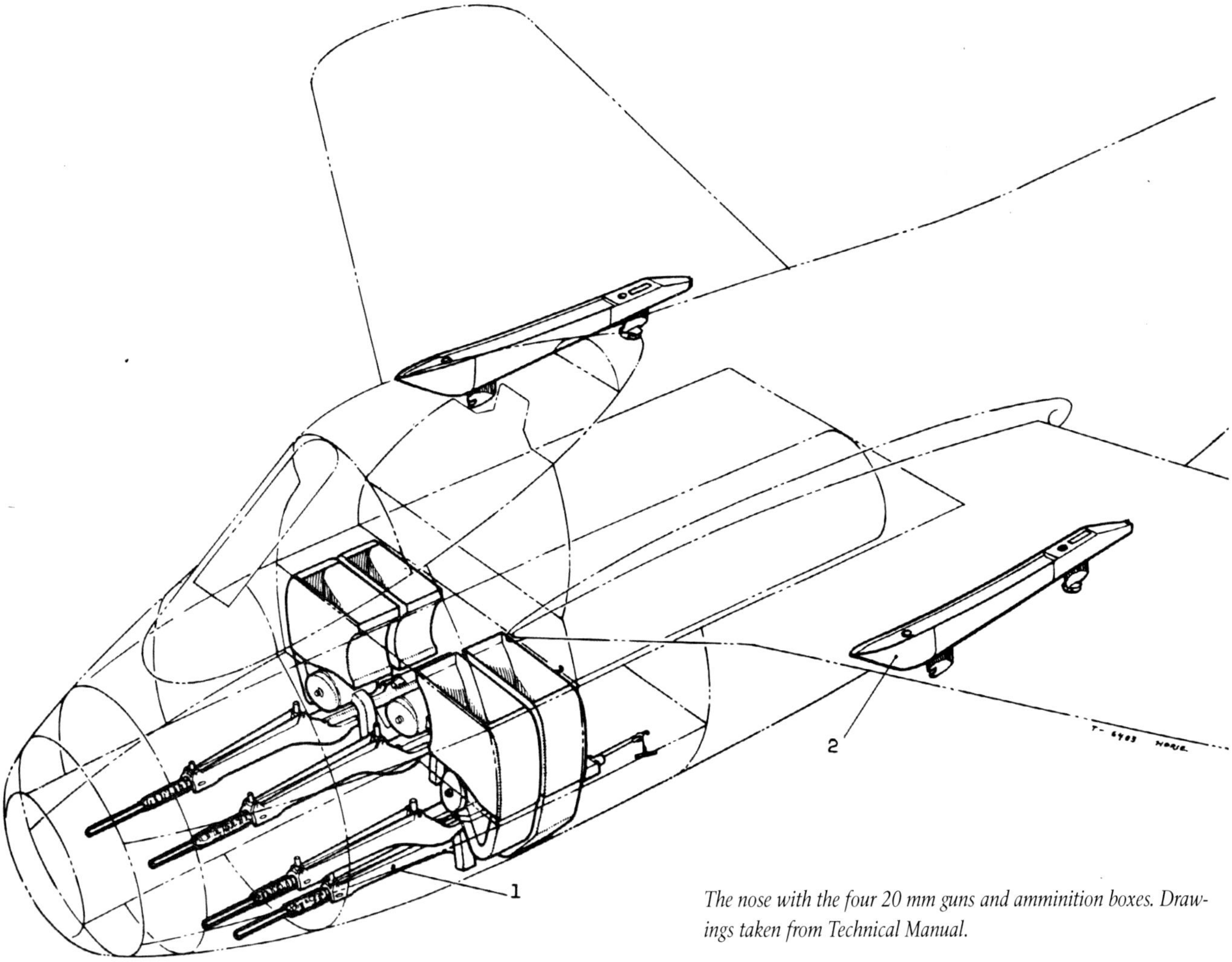

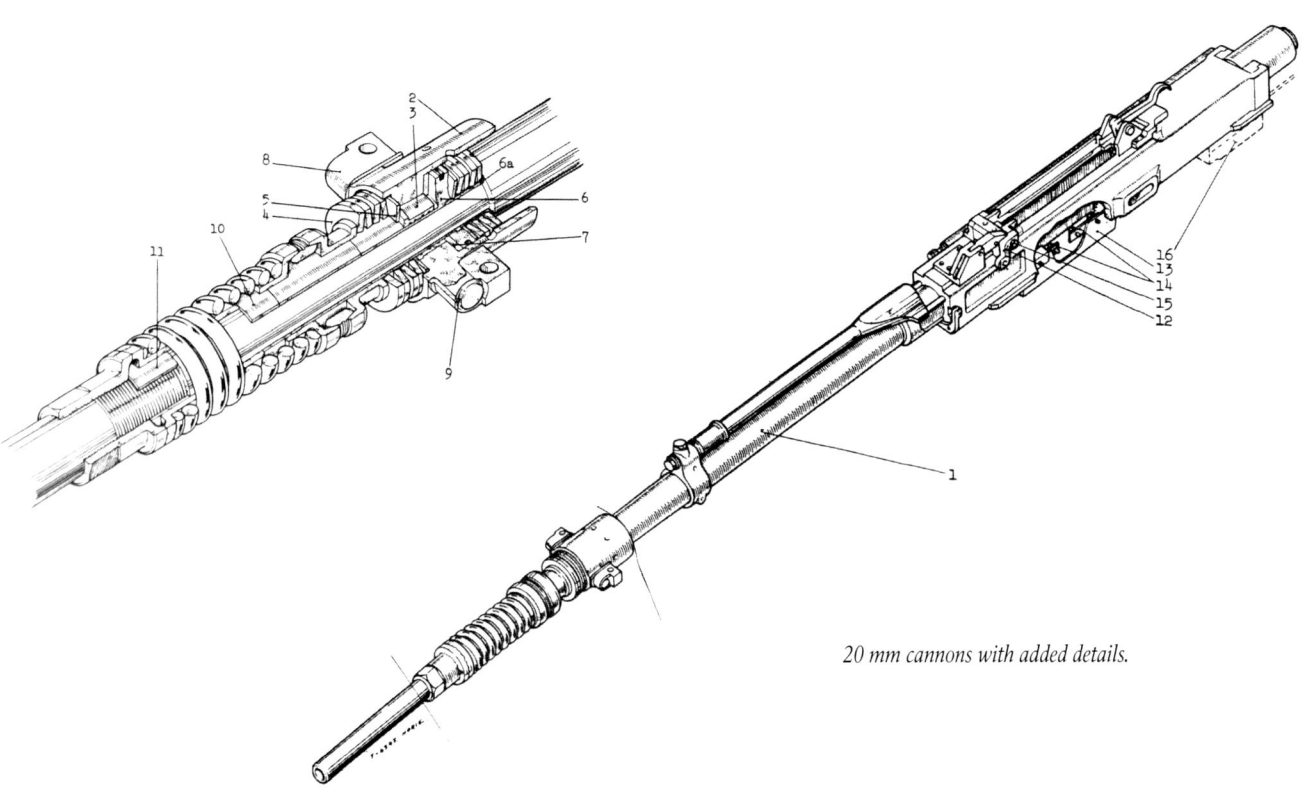

2

1

The nose with the four 20 mm guns and amminition boxes. Draw-
ings taken from Technical Manual.

8

2
3

5
4

6a

6

10

7

11

9

16
13
14
15
12

1

20 mm cannons with added details.

J 29Bs of Wing F 22: s/n 29475, code White J and s/n 29374, code White D. The former was lost on 10 January 1963, and the latter on 12 March 1963. (Archives of Svensk Flyghistorisk Förening)

J 29B, s/n 29475, code White J, under tow in Congo. (via Christer Ström)

SAAB J 29B, s/n 29425, White J, Wing F 22 Congo (Africa), 1962.
Camouflage colours: upper surfaces Dark Green (mörkgrön 326M) and Dark Earth (mörkbrun)
with Orange stripes, under surfaces natural metal (unpainted).

© Thierry Vallet / 2015

Thierry Vallet

SAAB J 29B, s/n 29425, White J, Wing F 22 Congo (Africa), 1962.

Thierry Vallet

SAAB J 29B, s/n 29425, White J, Wing F 22 Congo (Africa), 1962.

Thierry Vallet